Busy Ant Maths

2nd EDITION

Activity Book
Foundation B

Add

$$\boxed{1}\ \boxed{2}\ \boxed{3}\ \boxed{4}\ \boxed{5}\ \boxed{6}\ \boxed{7}\ \boxed{8}\ \boxed{9}\ \boxed{10}$$

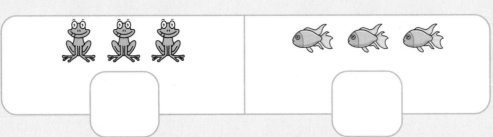

Teacher's notes

Write the number of frogs. Write the number of fish. Write the total.

2

Date: _____

Add

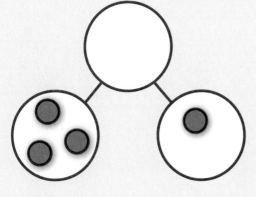

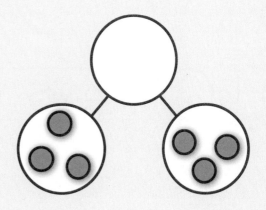

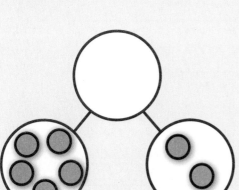

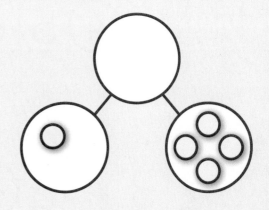

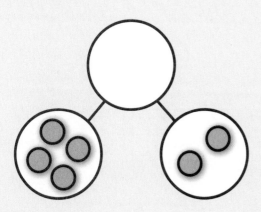

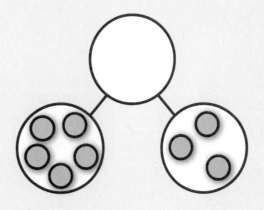

Teacher's notes

Complete each part-part-whole model: write the total number of counters in the 'whole' section.

3

Add on

Date: _____

1	②	3	4	5	6	7	8	9	10

1	2	3	④	5	6	7	8	9	10

1	2	3	4	5	⑥	7	8	9	10

1	2	3	4	⑤	6	7	8	9	10

Teacher's notes

Start with the circled number. Count on the number of fingers. Colour the total on the number track.

4

Add

Date: _____

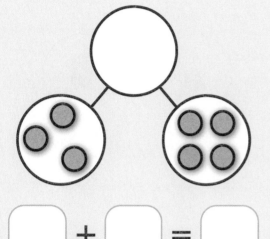

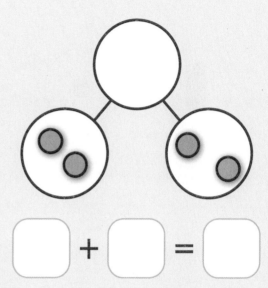

[ ] + [ ] = [ ]   [ ] + [ ] = [ ]

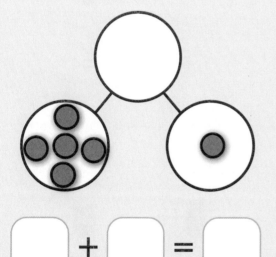

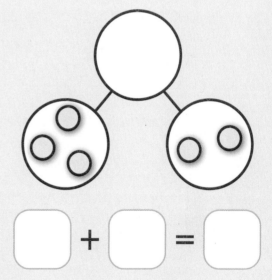

[ ] + [ ] = [ ]   [ ] + [ ] = [ ]

Date: _____

Totals to 4

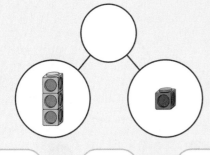

```
[ ]  +  [ ]  =  [ ]
```

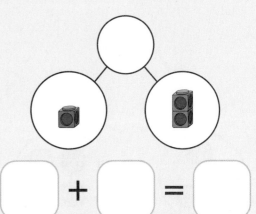

```
[ ]  +  [ ]  =  [ ]
```

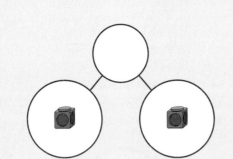

```
[ ]  +  [ ]  =  [ ]
```

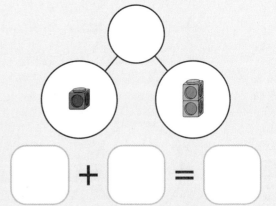

```
[ ]  +  [ ]  =  [ ]
```

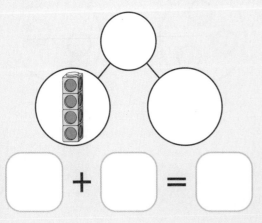

```
[ ]  +  [ ]  =  [ ]
```

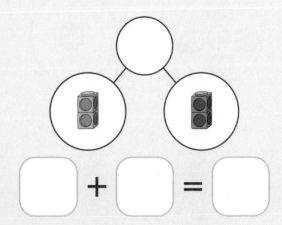

```
[ ]  +  [ ]  =  [ ]
```

Teacher's notes

Complete each part-part-whole model: write the total number of cubes in the 'whole' section.
Then write an addition number sentence to match.

6

Totals to 4

Date: _____

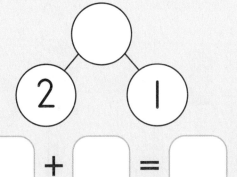

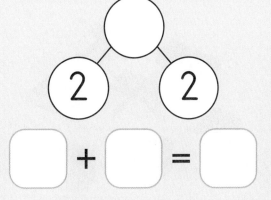

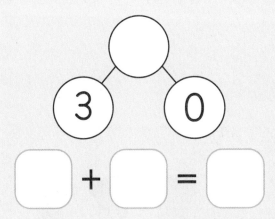

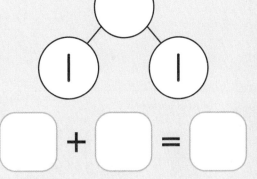

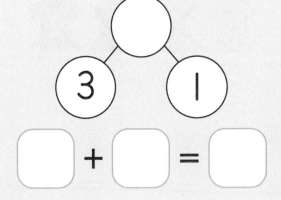

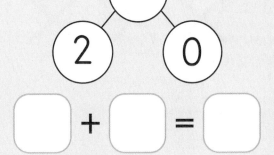

Date: _____

Take away

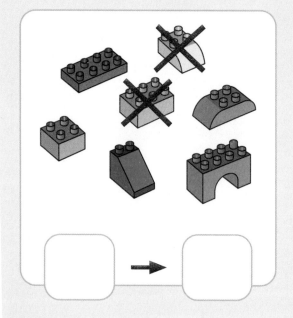

$\longrightarrow$

$\longrightarrow$

$\longrightarrow$

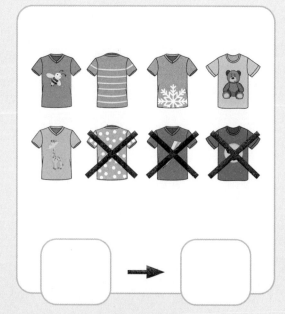

$\longrightarrow$

Teacher's notes

In each set, record how many there were to start with. Record how many are left.

Date: _____

Take away

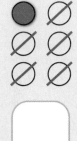

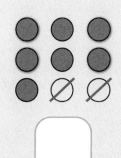

Teacher's notes

In each set, some counters have been taken away. Count how many counters are left. Write the number in the box.

9

Count back

Date: _____

Teacher's notes

Count the sticks. Circle that number on the track. Count back along the number track the number of balloons that have burst. Colour the number you reach on the track.

Date: _____

Count back

0	1	2	3	4	5	6	7	8	9	10

0	1	2	3	4	5	6	7	8	9	10

0	1	2	3	4	5	6	7	8	9	10

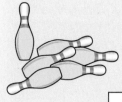

0	1	2	3	4	5	6	7	8	9	10

Teacher's notes

Count the skittles. Circle that number on the track. Count back along the number track the number of knocked-down skittles. Colour the number you reach on the track.

11

Subtract

Date: _____

☐ − ☐ = ☐   ☐ − ☐ = ☐

☐ − ☐ = ☐   ☐ − ☐ = ☐

Date: _____

Subtract from 2, 3 or 4

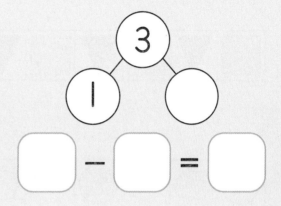

□ − □ = □

□ − □ = □

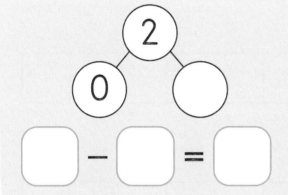

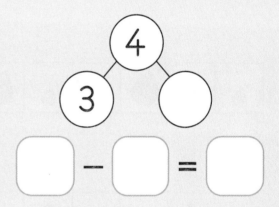

□ − □ = □

□ − □ = □

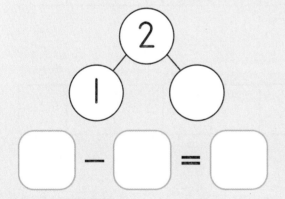

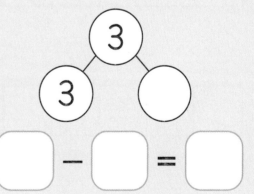

□ − □ = □

□ − □ = □

Teacher's notes

Complete each part-part-whole model: write the missing 'part' to show how many are left. Then write a subtraction number sentence to match.

Date: _____

Patterns

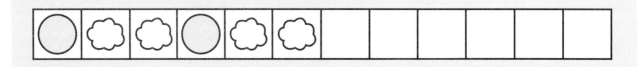

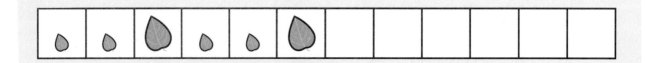

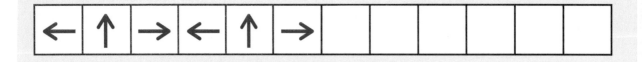

Teacher's notes

Continue each pattern. Then draw your own pattern.

Patterns

Date: _____

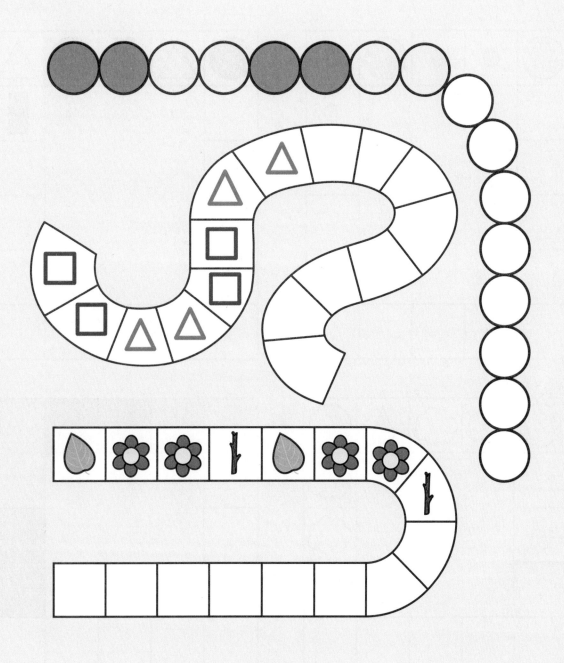

Patterns

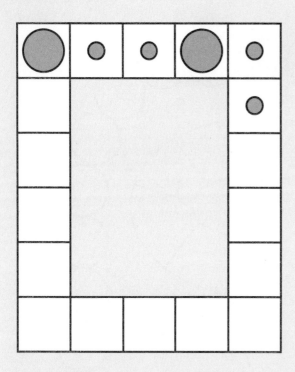

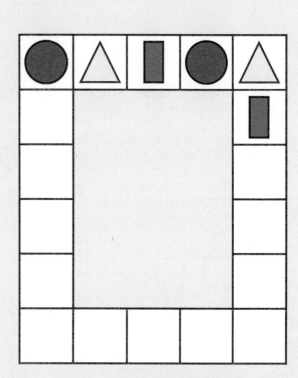

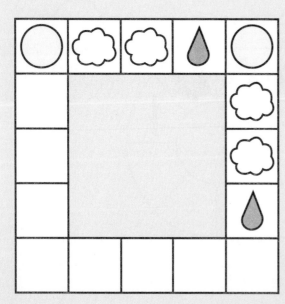

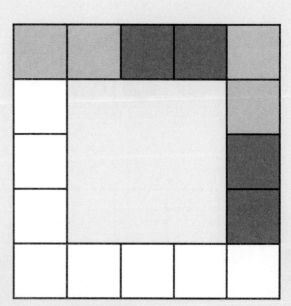

Teacher's notes

Complete each pattern.

Date: _____

Sort

blue ☐   **yellow** ☐

2 holes ☐   **4 holes** ☐

Doubles

Date: _____

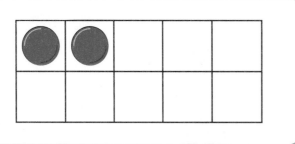

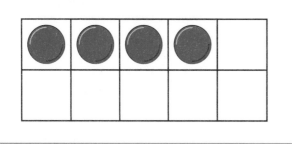

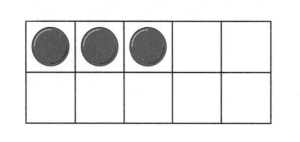

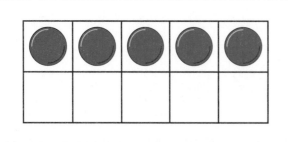

Teacher's notes

For each ten frame, count the number of red dots and write the number in the circle. Then draw the same number of dots in the second row. Finally, count all the dots and write the number in the box.

18

Equal groups

Date: _____

Teacher's notes

For the three shirts, draw the same number of buttons on each shirt. Write the total number of buttons in the box. For the four fish bowls, draw the same number of fish in each bowl. Write the total number of fish in the box.

Date: _____

Half

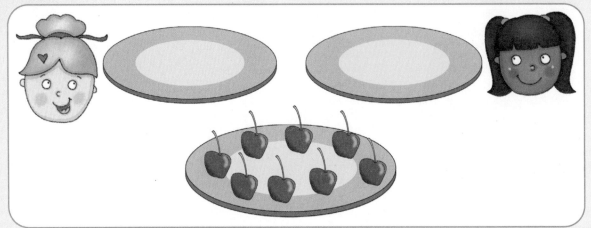

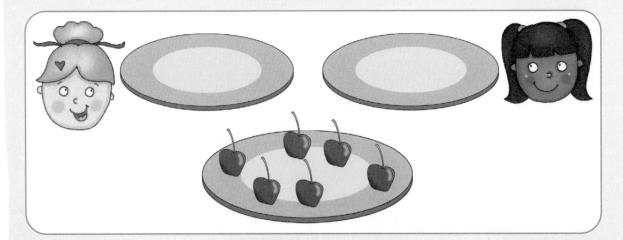

Teacher's notes

For each box, share the cherries equally by drawing on the small plates how many cherries each child gets.

Even and odd

Date: _____

| 0 | 1 | 2 | 3 | 4 | 5 | 6 | 7 | 8 | 9 | 10 |

| 0 | | | | | |

| 0 | 1 | 2 | 3 | 4 | 5 | 6 | 7 | 8 | 9 | 10 |

| 1 | | | | |

Teacher's notes

Continue the jumps counting on in twos. Then write all the numbers you land on in the small boxes underneath. In the large boxes, write any other even or odd numbers you know.

21

Today, yesterday, tomorrow

Date: _____

Monday

| Tuesday |

| Wednesday |

| Thursday |

| Friday |

| Saturday |

| Sunday |

yesterday

today

tomorrow

Teacher's notes

Draw a line to match the correct day to 'today'. Repeat for 'yesterday' and 'tomorrow'. Then draw
something you did/might do on each day in the boxes.

My day

Date: _____

O'clock

Date: _____

Teacher's notes

Look at the clock. Draw something you might do at that time of day.

24

Longer time

Date: _____

Teacher's notes

In each pair, tick the activity that takes a longer time.

Estimate, then count

Make numbers

Date: _____

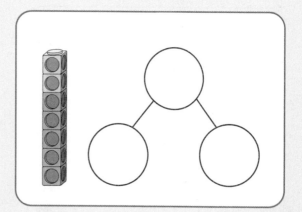

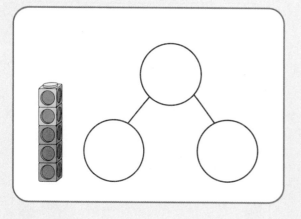

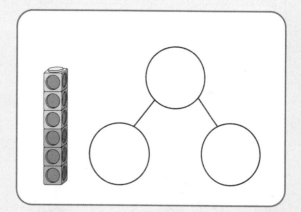

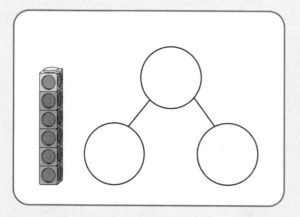

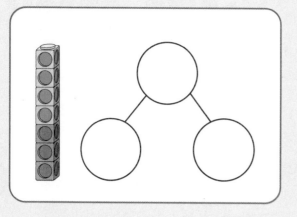

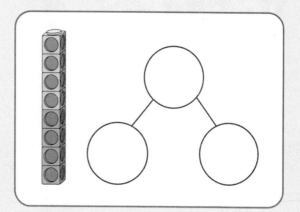

Teacher's notes

Complete each part-part-whole model to match the numbers of cubes.

27

Date: _____

More

Teacher's notes

Count the leaves on each tree. Write the number in the box. For each pair, circle the number that
is **more**.

Smallest and largest

Date: _____

Teacher's notes

For each strip of bunting, colour the **smallest** number **red**, and the **largest** number **blue**.

29

0

1

2

3

4

5

6

7

8

9

10

11

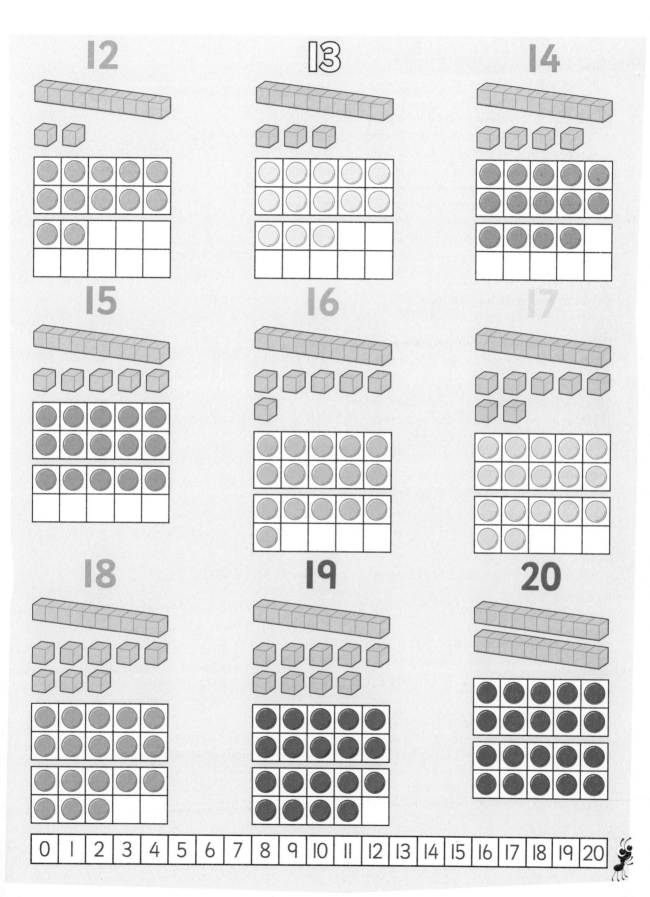

Assessment record

_____ has achieved these Maths Reception objectives:

Counting and understanding numbers

• Count on and back in ones, starting from any number from 0 to 20.	1	2	3
• Count objects, actions and sounds from 0 to 20.	1	2	3
• Recognise the number of objects presented in familiar patterns up to 10 without counting (subitise).	1	2	3
• Estimate a group of objects and check by counting.	1	2	3
• Compose (put parts together to make a whole) and decompose (break down a number into parts) numbers to 10.	1	2	3
• Explore and represent patterns within numbers up to 10, including evens and odds, double facts and how quantities can be distributed equally.	1	2	3

Reading and writing numbers

• Read and write numbers from 0 to 20.	1	2	3

Comparing and ordering numbers

• Understand the relative size of quantities to compare numbers from 0 to 20.	1	2	3
• Understand the relative size of quantities to order numbers from 0 to 20.	1	2	3

Understanding addition and subtraction

• Understand addition as combining two sets.	1	2	3
• Understand addition as counting on.	1	2	3
• Understand subtraction as take away.	1	2	3
• Understand subtraction as counting back.	1	2	3
• Find 1 more and 1 less than a number from 1 to 10.	1	2	3

Patterns and sequences

• Talk about, copy, continue and create repeating patterns.	1	2	3

Time

• Use familiar language to describe time, including year, month, week, day and hour.	1	2	3
• Sequence familiar events.	1	2	3
• Recognise the time to the hour.	1	2	3
• Compare intervals of time.	1	2	3

Statistics

• Sort and match objects, pictures and children themselves, explaining the decisions made.	1	2	3
• Count how many objects share a particular property.	1	2	3
• Present results using practical resources, pictures, drawings and numbers.	1	2	3
• Describe data using everyday language such as more, less, most or least.	1	2	3

1: Emerging   2: Expected   3: Exceeding

Signed by teacher:
Signed by parent: Date: